Health 96

蚊子的叮咬

Mosquito Bites

Gunter Pauli

[比] 冈特·鲍利 著

[哥伦] 凯瑟琳娜·巴赫 绘

何家振 译

上海远东出版社

丛书编委会

主　任：田成川

副主任：何家振　闫世东　林　玉

委　员：李原原　翟致信　靳增江　史国鹏　梁雅丽
　　　　任泽林　陈　卫　薛　梅　王　岢　郑循如
　　　　彭　勇　王梦雨

特别感谢以下热心人士对童书工作的支持：

匡志强　宋小华　解　东　厉　云　李　婧　庞英元
李　阳　刘　丹　冯家宝　熊彩虹　罗淑怡　旷　婉
杨　荣　刘学振　何圣霖　廖清州　谭燕宁　王　征
李　杰　韦小宏　欧　亮　陈强林　陈　果　寿颖慧
罗　佳　傅　俊　白永喆　戴　虹

目录

Contents

一匹斑马在水潭旁喝水，这时一只猴子停下来哀叹道："我昨晚睡不着，那些蚊子简直太讨厌了！我想在白天打个盹都不行，它们嗡嗡的声音不停地烦我！"

"真有意思！"斑马说，"蚊子从来不来烦我。"

A zebra is taking a long drink at the waterhole when a monkey stops by and laments, "I couldn't sleep last night, you know. Those mosquitoes were simply impossible! And I cannot even catch a nap during the day either. Their buzzing irritates me endlessly!"

"Interesting," says the zebra, "I have no mosquito problems."

那些蚊子简直太过厌了！

Those mosquitoes were simply impossible!

Black and white stripes

"那是因为你的皮肤很厚！"

"不，不是那样。我黑白相间的斑纹能驱赶蚊子。"

"我知道你黑白相间的条纹能让你凉快，但我还是头一回听说它能驱赶蚊子。"

"It's because your skin is so thick!"

"No, it's not that. My black and white stripes keep the mosquitoes away."

"I know your black and white stripes cool you down, but that it keeps mosquitoes away, now that is news to me."

"的确是这样，我身上有阵阵微风吹过，蚊子没法停在那里。"

"太聪明了！我也希望我有办法让蚊子不叮我。蚊子简直一无是处。"

"自然界里，任何事物都有存在的理由。"

"Sure, too many micro gusts of wind blow over my body, so they cannot land."

"That's smart! I wish I had a way of keeping mosquitoes away. They are good for nothing."

"Everything in nature has a reason to exist."

任何事物都有存在的理由

Everything has a reason to exist

老兄，在蚊子吸过我的血之后，我的确感到痒！

Boy, it does itch after they sucked my blood!

"是的，我知道。但我需要绞尽脑汁才能想到蚊子在这个世界上究竟有什么益处。我甚至在想，也许蚊子能提供微按摩。"

"天晓得。当蚊子叮你的时候你觉得疼吗？"

"不，一点都不疼。不过，老兄，在蚊子吸过我的血之后，我的确感到痒。"

"Yes, I know. But I've been racking my brain to try to imagine what good mosquitoes could do in the world. Once I thought perhaps mosquitoes offer tiny massages."

"Who knows? But have you ever felt pain when a mosquito stings you?"

"No, nothing. But, boy, it does itch after they sucked my blood!"

"你胳膊上打过针吗？"

"当然打过啊，现在根本没法逃过接种疫苗，不管你喜不喜欢。"

"你害怕打针吗？"

"害怕？我一看到针就晕了。只是想到打针，就会让我感到虚弱。"

"And have you ever had an injection in your arm?"
"Sure, there is no way to escape vaccinations these days, whether you like it or not."
"Are you scared of needles?"
"Scared? I nearly faint when I see one. Just the thought of it makes me weak."

你害怕打针吗?

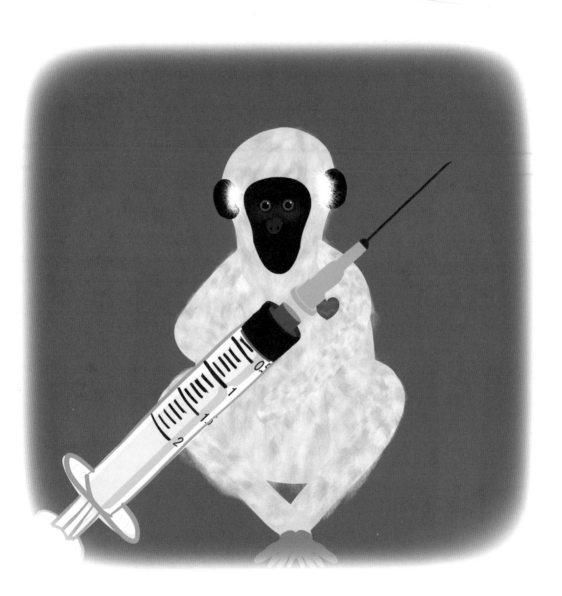

Are you scared of needles?

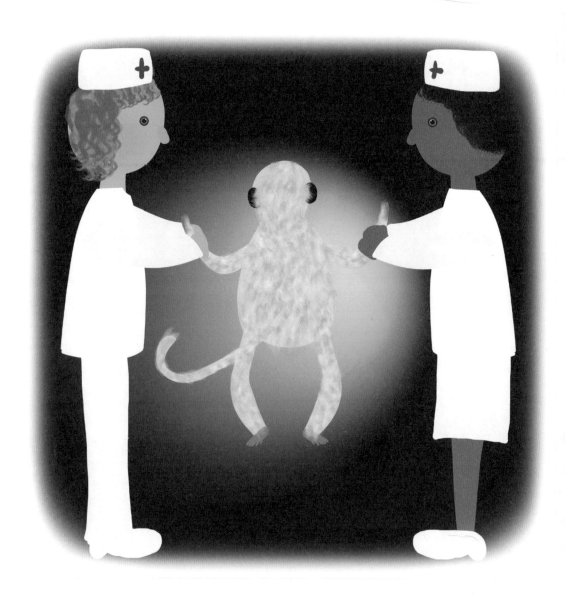

Two nurses had to hold me down

"打针时你觉得疼吗？"

"哦，是的，很疼。有一次我还跳了起来，结果不小心被针扎破了，两名护士不得不摁住我。我妈妈觉得很难为情。"

"你想过吗，为什么蚊子咬你的时候你感觉不到，但是打针的时候却觉得疼呢？"

"And does it hurt when you get an injection?"

"Oh yes, it hurts. Once I jumped up and was accidentally scratched by a needle. Two nurses had to hold me down. My mum was so embarrassed."

"Have you ever wondered why you do not feel it when a mosquito bites you, but you do feel pain when a needle goes into your arm?"

"因为注射针更粗？"

"最细的针头几乎和蚊子的喙一样细。"

"是不是蚊子用了止痛药，让我感觉不到痛？"

"不，那是由蚊子喙的形状决定的。这完全是几何结构的问题。"

"The needle is thicker?"

"The finest needles are nearly the same size as a mosquito's proboscis."

"Mosquitoes use a painkiller so I feel nothing?"

"No, it's because of the shape of the proboscis. So it's all about geometry."

这完全是几何结构的问题

It's all about geometry

圆锥形的，而且还有锯齿状的表面

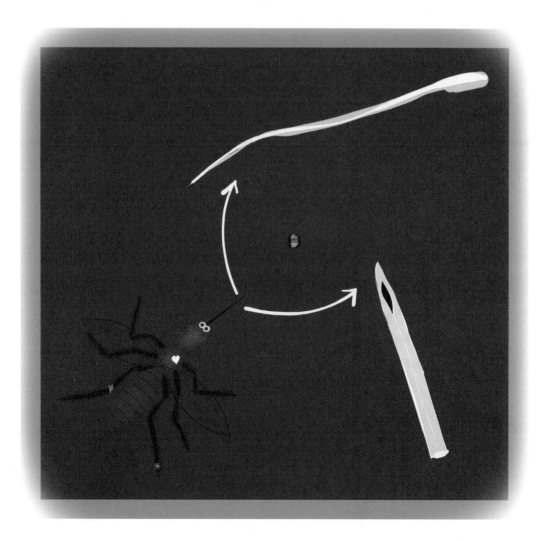

shape of a cone and surface like a saw

"你是想显摆你很聪明吗？你是想聊聊数学之类的？我在数学考试时可痛苦了。"

"告诉我，蚊子的喙是什么形状的？"

"你是说它的长鼻子吗？让我想想……圆柱形？"

"不对，喙是圆锥形的，而且还有锯齿状的表面。"

"You want to be smart and talk about mathematics and so on? You know I had a hard time passing those exams."

"Tell me, what shape is a mosquito's proboscis?"

"You mean its long snout? Let me think … a cylinder?"

"No, it has the shape of a cone and has a surface like a saw."

"这是为了扎进我的皮肤？"

"不，是为了确保你什么都感觉不到！"

"为什么医生不把注射器做成那样呢？这样我妈妈再也不会感到难堪了！"

……这仅仅是开始！……

"Cutting into me?"

"No, making sure you feel nothing!"

"So why do not the doctors make use of syringes like that? My mum will never have to be embarrassed again!"

... AND IT HAS ONLY JUST BEGUN!...

……这仅仅是开始！……

… AND IT HAS ONLY JUST BEGUN! …

Did You Know?

你知道吗?

斑马黑白相间的斑纹，使其皮肤表面有温差并产生了空气流动，让它们保持凉爽。在皮肤表面吹过的风还可以阻止蚊子叮在它们身上。

Zebras' black and white stripes create temperature differences and wind on their hides, keeping them cool. Wind blowing over their skins prevents mosquitoes from landing.

蚊子并不叮或者咬，而是不断扭动它们的喙来探寻你身上的血管。蚊子没有坚硬锋利的针头；它们的喙更像一个带圆锥尖的触须。

Mosquitoes do not sting or bite, they probe by twisting and bending their mouth parts (proboscis) in search of a blood vessel in your skin. A mosquito does not have a sharp and rigid "needle"; its "snout" rather operates like a tentacle with a conical end.

蚊子的喙尖是圆锥形的，这就是为什么它刺进你的皮肤时，你什么都感觉不到。

The shape of the top of the mosquito's proboscis is conical, which explains why you do not feel anything when the mosquito probes.

"蚊子"这个词来自西班牙语，意思是"小飞虫"。蚊子传播疟疾、黄热病、登革热。蚊子有冬眠期或者滞育期，在太冷或者湿度太低时停止发育。

The word "mosquito" is borrowed from Spanish, which means "little fly". Mosquitoes transmit malaria, yellow fever, and dengue fever. Mosquitoes hibernate, or diapause, delaying their development when the weather is too cold and the humidity too low.

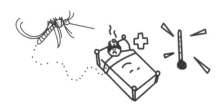

雄性蚊子能活 5 到 7 天，以花蜜和糖为食，给花儿授粉。雌性蚊子为了蚊卵的发育，必须吸食充足的血液。产卵后，它又开始寻找更多的血液。

Male mosquitoes live five to seven days, feeding on nectar and sugar, and pollinating flowers. The female searches for a full blood meal, which is needed to develop her eggs. Once the eggs have been laid, she starts searching for more blood.

蚊子是地球上最致命的昆虫，尤其是当它携带了疟疾病菌时。此外，作为水生昆虫，蚊子的幼虫是食物链中重要的一环。蚊子幼虫是鱼类的营养快餐。成年蚊子为鸟类提供了大量的生物质食物。

Mosquitoes are the deadliest insects on Earth, especially as malaria vectors. Still, as aquatic insects, mosquito larvae play a key role in the food chain. Mosquito larvae are nutrient-packed snacks for fish. As adults, they represent a considerable biomass of food for birds.

The estimated ratio of insects to humans is 200 million to one. There are 160 million insects per hectare of land.

据估计，昆虫与人类数量的比例是 2 亿比 1。1 公顷土地上有 1.6 亿只昆虫。

A million people die worldwide per year due to unclean syringes. The disposable syringe is only a temporary solution. Needle-free injectors are a much more sustainable solution but rely on power.

全球每年有 100 万人死于不干净的注射。一次性注射器只是临时的解决方案，无针注射才是更长久的解决方案，但是依赖于电力。

想象一下，如果昆虫是地主，人类是佃农，谁活下来的机会更大？

Imagine insects are the landowners and we are the tenants. Who has the greatest chance of survival?

你愿意了解更多蚊子在生态系统中的积极用途，还是愿意学习如何消灭蚊子？

Would you like to learn more about the positive role of mosquitoes in ecosystems, or would you rather want to learn how to kill them?

如果注射器针头的形状能够消除疼痛，恐惧打针的人会减少吗？

If the shape of a needle eliminates pain, would less people have a phobia for needles?

你相信自然界里的一切都有其存在的理由吗？

Do you think that everything in nature has a reason to exist?

You are in charge of an immunisation campaign. Children are scared of needles; even some of their parents are scared. How do you overcome this fear? Would you impose strong rules and control, talk about the benefits of getting the vaccine, have a smiling team to accompany doctors, buy extra thin needles that give less painful injections, have needle-free injection systems, tell funny stories while injecting to distract the patients, or play noisy TV cartoons? Discuss your ideas with your friends and come up with some strategies. Perhaps you have different strategies you would like to share with us, ones for children of different ages or even a special one for adults who are scared of needles.

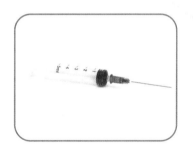

假设你在负责一项免疫接种活动。孩子们害怕打针，甚至有些父母也害怕。你如何帮助他们克服这种恐惧？强行控制？宣传接种疫苗的好处？组建微笑团队配合医生？购买超细针头以减轻注射时的疼痛？使用无针注射器？在打针时讲一些搞笑的故事或者播放喧闹的电视卡通来分散孩子的注意力？与你的朋友讨论，想出一些办法。也许你可以和我们分享一些不同的办法，包括适合不同年龄段孩子甚至是专门针对害怕打针的成年人的办法。

学科知识

Academic Knowledge

生物学	一次按摩可以减轻焦虑，降低血压和心率，持续性的按摩治疗可以减轻特质性焦虑、抑郁和疼痛；蚊子喙可以作为探测血管的触角；已知的蚊子有3 500种；引进食蚊鱼的生物控制方法已在南美和黑海附近的俄罗斯南部成功减少疟疾，但是在另外一些地方，食蚊鱼对当地鱼类造成了损害；蚊子能通过翅膀振动的声音辨识其家庭成员；蚊子的耳朵能听到并分辨不同的嗡嗡声，并且能够通过变声传达特殊信息；蚊子在静止的水中产卵，包括湖水和肮脏的死水。
化　学	按摩有助于释放身体里的化学物质；蚊子的唾液会刺激人的皮肤；蚊子唾液是一种抗阻剂，能防止喙在刺入皮肤时被阻塞；二氧化碳、辛醇、汗水以及人类身体发出的红外线对蚊子有吸引力。
物　理	利用白色白天反射热量，夜晚保存热量；利用黑色白天吸收热量，夜晚散发热量；黎明和傍晚，蚊子会在无风的环境里叮咬人；叮意味着注射毒液，咬意味着吸血；雌性蚊子体型更大一些，因此它们的翅膀扇动得更慢一些。
工程学	无针注射器通过喷嘴喷出高速液体射穿皮肤，避免了因不正确的针头消毒方法带来的问题；蚊子通过化学物质、视觉和热导感受器发现猎物。
经济学	治疗疟疾的成本使非洲经济增长率下降了约1.3%。
伦理学	自然界没有"好"和"坏"，万物皆有存在的理由；人类决定消灭某些物种（例如蚊子）是一种自大的表现；引进非本地物种以解决某一问题往往会引发很多新的问题。
历　史	世界上最古老的蚊子是在7 900万年前的加拿大琥珀中发现的。
地　理	除了南极和冰岛外，世界各地都有蚊子。
数　学	数学模型有助于预测流行病暴发的趋势，评估蚊子控制技术的实效。
生活方式	每年因注射针刺伤而意外死亡的人数超过100万；在影片《侏罗纪公园》中，科学家通过抽取200万年前曾经吸食过恐龙血的蚊子身上的基因，克隆出恐龙。
社会学	共同威胁（如蚊子）意识动员人们团结起来应对威胁；会嗡嗡叫的蚊子不咬人的说法是对的，因为只要它在发出声响，它一定是在飞行，而在飞行中的蚊子是不可能咬人的。
心理学	恐惧症是对一种真实威胁的非理性恐惧反应。
系统论	蚊子对人的健康是有害的，但同时，它们的生物量是巨大的，它们在生态系统中的作用是明确的，它们为鱼类和鸟类提供营养；一物种对另一物种的害处，并不是消灭该物种的理由，因为那样将打破生态平衡，结果将事与愿违。

情感智慧
Emotional Intelligence

斑 马

斑马很放松。他知道自己防御蚊子的机理，并且很乐意分享他的知识，尽管这并不能解决猴子的烦恼。斑马站在哲学立场上，认为蚊子是一种讨厌的东西。他相信万物皆有存在的理由。斑马通过一连串的问话，了解到猴子害怕蚊子和打针。这显示了斑马的态度和求知欲，表明他的思维很清晰。他有一个高尚的目标：消灭痛苦。

猴 子

猴子很紧张，睡不着觉。他渴望学习，但是缺乏知识、洞察力和想象力。他的思路不畅，想象不出蚊子对生态系统的益处。斑马的问题揭示了猴子恐惧打针的事实。猴子清醒而诚实，说出了他对蚊子的厌恶。他坦承希望找到克服非理性恐惧的办法，这并不丢面子。恐惧使他和他的家人感到难堪。这些对话使猴子更加自信，并在猴子和斑马之间建立了同理心。

艺术
The Arts

蚊子具有非常与众不同的嗡鸣声。我们可以模仿它。我们可以开一个蚊子嗡鸣声的音乐会，除了我们的嘴巴，不使用任何乐器。你知道一首你和你的小伙伴都能嗡嗡哼唱的歌吗？

思维拓展
Systems: Making the Connections

全世界每年有2.5亿人感染疟疾，100万人因疟疾死亡。因此，蚊子给人类带来沉重的医疗和财政负担。很多人会不假思索地认为应该消灭蚊子。然而，这种想法会带来一些问题。首先，只有10%的蚊子是咬人或者烦人的。这种昆虫在地球存在了上亿年，已经成为生态系统的一部分。人们并没有充分认识到蚊子在生态系统中扮演的角色以及它们作出的贡献。没有蚊子，蚊子的天敌就没有食物，某些植物就失去了为其授粉的昆虫。消灭蚊子将在北极地区造成最大的生态差异。在一年中很短的一段时间内，北极地区的蚊子数量异常多，形成一团团厚厚的蚊子云。它们是食物源，如果没有蚊子，候鸟的数量将减少50%。北美驯鹿迎着风逃离蚊子的烦扰，而逃离的路径决定了北美驯鹿的迁徙路线。如果没有蚊子的幼虫，数百种鱼将不得不改变它们的饮食，受影响的可不仅仅是灭蚊高手食蚊鱼（它们常被引进稻田或池塘里以灭蚊）。在安第斯山脉的热带地区，蚊子是可可树的传粉者。如果没有蚊子，我们将会为吃巧克力付更多的钱。消灭一个物种也许能暂时缓解人类的痛苦，但此物种将很快被其他物种取代，而这些新物种很有可能也会给人类带来多种疾病。人类无意间将一些有益的物种逼到濒临灭绝的边缘，比如金枪鱼、珊瑚，他们现在又想消灭蚊子。但是蚊子在自然界占据着至关重要的地位。我们必须认识到，如果蚊子被成功消灭，生态系统可能只是会遇到一些小问题，然后继续发展，但我们又会遇到其他东西，可能更好，也可能更坏。

动手能力
Capacity to Implement

发明一个捕蚊器。目标是让蚊子不能逃走，但也不要杀死它们。只是捉到它们，然后用捉到的蚊子喂鱼。你要做的第一件事，是要想一想什么东西可以吸引蚊子：什么食物、气味和音乐。接着，你要找到什么能驱赶它们，换句话说，怎样做会捉不到蚊子。列出一个吸引蚊子的物品清单，并找到现成的、低成本的东西。假如你找到了一种行得通的方式，你将很快成为一名重要的社会企业家，因为每个人都想在家里有一个捕蚊器。

故事灵感来自
This Fable Is Inspired by

冈野雅行
Masayuki Okano

冈野雅行在 16 岁时从初中退学，加入了他父亲的公司，制造模具和压模。作为匠人的父亲认为，儿子是一个变不可能为可能的梦想家。冈野雅行发明了有更大开口的易拉罐。他相信工作上的成功不仅与能力有关，也与良好的人际关系有关。他后来又发明了无痛胰岛素注射针，直径只有接种疫苗针头的一半。这种针头直径只有 0.2 毫米，针尖是像蚊子的喙一样的圆锥形。虽然这种设计想象起来相当容易，但是如何将其规模化生产是个巨大的挑战。泰尔茂公司认为这是不可能的，但冈野雅行解决了这个问题。他的发明将数百万需要定期打针的糖尿病人从痛苦中解脱出来。

图书在版编目（CIP）数据

冈特生态童书.第三辑修订版:全36册:汉英对照 /
（比）冈特·鲍利著;（哥伦）凯瑟琳娜·巴赫绘;
何家振等译.—上海:上海远东出版社,2022
书名原文:Gunter's Fables
ISBN 978-7-5476-1850-9

Ⅰ.①冈… Ⅱ.①冈… ②凯… ③何… Ⅲ.①生态环
境-环境保护-儿童读物—汉、英 Ⅳ.①X171.1-49

中国版本图书馆CIP数据核字（2022）第163904号
著作权合同登记号图字09-2022-0637号

策　　划　张　蓉
责任编辑　程云琦
封面设计　魏　来　李　廉

冈特生态童书
蚊子的叮咬
[比]冈特·鲍利　著
[哥伦]凯瑟琳娜·巴赫　绘

何家振　译

记得要和身边的小朋友分享环保知识哦！
八喜冰淇淋祝你成为环保小使者！